The Universe Explained
to my Grandchildren

Hubert REEVES

THE UNIVERSE EXPLAINED
TO MY GRANDCHILDREN

Translated from the French by
Donald Winkler

SALAMMBO
PRESS LIMITED

First published in the United Kingdom in 2012 by

Salammbo Press
39A Belsize Avenue
London NW3 4BN

Originally published in French as
l'Univers expliqué à mes petits-enfants
by Les Éditions du Seuil, Paris
Copyright © Éditiond du Seuil, 2011

A CIP catalogue record for this book is available from the British
Library.

ISBN 9780956808226

Prelude

The title of this book brings to mind Victor Hugo's *How to be a Grandfather*. In a world where marriages are not as durable as they once were, grandparents have become important reference points, all the more so in that they tend to share the lives of their grandchildren for longer periods of time.

I have height grandchildren: Raphaëlle, Dorian, Elsa, Cyprien, Massis, and Noé, from 6 months to 21 years old. This book is dedicated to them. When I began to write it, I became aware of the symbolic value it could have: that of a spiritual testament.

What would I want to tell my grandchildren about the story of this great universe that will continue to be their home after I am gone? How might I help them to pass on, themselves, this knowledge?

I have chosen to address children who are about fourteen years old. And along with them, of course, all those who want to learn more about our cosmos and its history.

The book was born from conversations I had with one of my granddaughters, on certain summer evenings. Our dialogues took place under the starry sky, which we observed from the comfortable vantage point of our chaises-longues. All during its writing, I was able to relive those August evenings when the children showered me with questions while we awaited the shooting stars.

The contemplation of the heavenly vault and the sense of our presence among the stars inspired a shared wish to know more about this mysterious cosmos in which we live.

Here we will be talking about science, but that in no way rules out poetry.

An evening of observation

"Grandpa, when I told my friends we were going to write this book about the universe together , they had a lot of questions to ask me."

"For example?"

"How big is the universe? What was there before the Big Bang? Will there be an end to the world? How will it happen? And then: Are there other planets with people on them? Do you believe in extraterrestrials? My friends also say that in your books you compare a lot of things to food. They talk about alphabet soup and the raisin pudding your mother made."

"We'll talk about all of that. Thanks to science, and to astronomy in particular, we now understand many things. But there are many questions that have no answers. And many mysteries that have not been solved. I'm going to tell you about them so you won't begin to think that we know everything. Our universe is still extremely mystifying. Stretch out on your deckchair and close your eyes. Breathe deeply. Pay attention to all the parts of your body: your feet, your hands, your fingers. Your eyes, your ears, your nose. Are you ready?"

"Yes, I feel my whole body."

"For all of us, the universe begins with that:

what you feel, what makes it possible for you to see, to hear, to perceive both your inner world and the outside world. You are part of the universe, and it is through your body and your mind that we are going to explore it. Now open your eyes. It's night, the sky is clear. There are stars everywhere, some bright and others very dim, barely visible to the naked eye. There is the Earth that holds us up, the sun that brings light to the day, and the pale moon.

The universe is all of that. All, all, all.
But to begin, tell me how old you are."

"I'm almost fourteen."
"Where were you twenty years ago?"

"I didn't exist, grandpa!"
"Of course! I existed, but you didn't. Then something extraordinary happened. You were born, you came into the world, you began to be. You entered the universe. Before, you weren't there. I'm not talking about the day of your birth, which is now your birthday. I'm talking about the moment, about nine months earlier, when your mother and father conceived you while making love. That day is much more important for you than your birthday. On that day you appeared on a small planet, the Earth, which turns around the sun, which itself turns around the centre of our galaxy, the Milky Way, one of the countless galaxies in our universe. It all happened in your mother's womb. Millions of tiny cells with long tails

(spermatozoa) were injected by your father. There they embarked on a race. They surged towards the ovum that was lying in wait, and that would become your other half. What a tightly fought contest! But of all the contenders only one is important to us, the one that won the race. The one that penetrated the ovum and fertilized it. The others all died. And you, you came into existence thanks to those two cells that, in uniting, became you. You are now an inhabitant of the cosmos. It was at that moment that you began to live the long adventure of your life. During the nine months that followed, the little fertilized ovum became an embryo, then a foetus. Your body's cells arranged themselves in such a way that you were made ready to live in the world you entered into, and to gain knowledge of that world when you emerged from your mother's womb. Later, you opened your eyes, you looked upon the world, and you began to ask me questions such as: 'Grandfather, what is the universe?'

But already I have something astonishing to tell you: if there had not been, long before your birth, stars in the sky, you would not exist, you would never have been born. Me neither, for that matter. And we would not be here talking to each other."

"I can't imagine how the stars so far away in the sky could have anything to do with my existence. It's marvellous! How do you know that?"

"We'll get to that. But first, I'm going to explain to you many things."

How far are the stars?

"I'll never look at the stars the same way again. But I don't know how to tell if they're close by or far away. Tell me, how can you know, for instance, the distance between the Earth and the sun?"

"We'll begin by studying the sun. Tonight we'll go to our observatory to watch it set. This great luminous ball that slowly sinks towards the horizon is a star just like those we see at night. But the other stars are so far away that they seem very dim by comparison. Among all the stars in the sky, we are lucky enough to have one very close to us!"

"Yes, but how far away?"

"Obviously, it's farther than the mountains behind which it's going to set."

"Much farther?"

"Mankind asked itself this question for a very long time before finding the answer. Some people said it was very far away, others that it was very near. It was said that a prisoner, called Icarus, and his father, had planned to escape by rising into the sky with the help of two wings fixed to their backs with wax. But Icarus made a fatal mistake by flying too close to the sun: the wax melted, and he drowned in the ocean."

"So how can you measure the distance?"

"There are many ways. Here's one, for example, that works for the moon and the solar system. Remember our walks in the mountains last year? We had fun shouting to hear the echoes of our voices. Depending on the distance, they came back to us after a shorter or longer time. Sound (our shouting) travels fast: three hundred metres per second. If the echo comes back after two seconds (- one - two), you know that the cliff is three hundred metres away (one second to go, one second to return). To measure the distances in the solar system, we use the same method, not with sound, like for an echo in the mountains, but with light."

"There are echoes of light?"

"Yes, just like there are echoes of sound. But much faster: light travels a million times more quickly than sound. Today, to measure the distance to the moon, we send a laser flash (a kind of light) towards its surface. The echo comes back from the moon in two seconds (one to go, one to return). The moon is one light-second away.

To go to the sun, light takes eight minutes. We say that the sun is eight light-minutes away. There are sometimes violent storms that erupt there, with fiery bolts of lightning criss-crossing its surface. But we only see them eight minutes later. When we observe them from the Earth, we know that they were there eight minutes earlier. Why? Because those lightning flashes

12

had to travel the distance between the sun and ourselves."

"You mean the sun we're looking at tonight is the sun of eight minutes ago? How is it now? Has it changed in eight minutes?"

"To know that, we'll have to wait... eight minutes. In fact, we are just the right distance from our star. If it were farther away, it would be very cold, and we would not be able to live. If it were closer, it would be too hot, and the sea water would evaporate. Without that liquid, there would be no life at all. It's because our Earth is just the right distance from the sun that life has been able to develop and we are able to live comfortably.

Now let's wait for the night. The sun has set. The stars have appeared in the sky. Their light has travelled for a long time before reaching the Earth. Some stars that we see are dozens, hundreds, even thousands of light-years away. For example, the North Star, the one that points us in the direction north, is four hundred and thirty light-years away. For it to arrive today, its light would have left the star around the year 1580."

"And the three stars that you called the Belt, in the constellation Orion, how far away are they?"

"Their light has travelled one thousand five hundred years before reaching our eyes. It left towards the end of the Roman Empire, travelled all during the

Middle Ages, the Renaissance, and recent times, it flew through space, and finally got here. Of course, we can't measure those distances using echoes. We'd have to wait three thousand years for the round trip! Instead, we use other methods. You can read about them in astronomy books.

And now, if you look at the pictures of the cosmos taken with giant telescopes, you see a multitude of galaxies. Here, the distances are even greater. The light from some of them was emitted long before the birth of the Earth and the sun. It has been travelling almost since the beginning of the universe."

"How can we know what has become of them? Perhaps they don't even exist any more."

"That's a good question. We think that many of them have been swallowed up by larger ones. There is a lot of cannibalism among galaxies. But to know for sure, we'd have to wait for billions of years. Remember this well: when you observe a distant star, you are seeing it as it was in the distant past, and not as it is today. We can sum this up in the following way: 'To look far, is to look early.' Astronomers have at their beck and call a 'time machine' that those who study the history of the world can only dream of. They give us a direct view of the cosmos's past. For example, to know what the universe was like when the sun was born, 4.5 billion years ago, all we have to do is to observe stars that are 4.5 billion light-years away. That is what astronomers do today with their powerful telescopes.

That is how they reconstitute the history of the universe."

What are stars made of?

"You said that stars are very far away, but that they have played an important role in our life here on Earth. All I see are little points of light. How can we know what they are made of? And how could they have made a difference to our life here?"

"To answer your question, I'm going to go back to some ideas that are perhaps already familiar to you. We're going to talk about atoms and light."

"Yes, I've learned something about that, but I don't understand very well. Tell me about it as though I didn't know anything at all."

"Fine. Let's start at the beginning. Look around you. You see many different substances: the earth and stones that make up the ground on which you walk, the water you drink, the air you breathe, your food   fruits and vegetables. And also your body that you feel. One of the great discoveries of science was to show that all these substances, so numerous and so different, are in fact combinations of small particles we call atoms. They bear names that are familiar to you: oxygen, carbon, iron, chlorine, sodium, hydrogen, helium, lead, gold, etc. There are about a hundred of them. Here are some examples: water is made up of hydrogen and oxygen, table salt is chlorine and sodium, stones consist mainly of oxygen, silicon, iron,

17

and magnesium. Your body is primarily oxygen, carbon, nitrogen, and hydrogen. The air you breathe is above all a mixture of oxygen and nitrogen. This idea that the substances we perceive are combinations of atoms is already two thousand years old. It was proposed by Greek philosophers such as Democritus and Lucretius. But it's only in the eighteenth and nineteenth centuries that chemists were able to demonstrate its validity."

"All of that is on the Earth. Is it the same for the stars and planets? How can we know whether the sun is made of atoms the same way we are? It's so far away, and atoms are so tiny!"

"For me to give you an answer we must now talk about light and colours. Let's begin with the fluorescent lights used in advertising signs. There is, of course, the red of Coca-Cola emitted by atoms of hydrogen enclosed in glass tubing. There is also the yellow of the lamps that illuminate tunnels along the road, and the violet of mercury vapour lamps."

"How do we make these atoms give off light?"

"When we provide an atom with energy, for example through an electric charge, it gets rid of the energy by giving off light. Hydrogen lights up mainly in red, sodium in yellow, mercury in violet. These colours are a kind of signature that enables us to identify what kind of atom it is. And that can be anywhere: on Earth or in space, and even on the edge of the universe."

18

"So it's by observing the colours of the stars that we can know what they're made of? That's brilliant! Who had that idea?"

"It was a German astronomer, Joseph von Fraunhofer, who made the first analysis of sunlight in 1811. There he found the signatures of a number of different atoms: hydrogen, calcium and so on. And so the sun, like us, is made up of atoms. So are the planets, and all the stars we see in the universe. We find there every atom that we know. And only those that we know. We have not observed in the sky any atoms unknown on Earth. Are you getting an idea of how important this discovery was? Thanks to the colour of the light we see through our telescopes, we can know the atomic composition of everything that shines in the sky!

Just as a footnote, around the same time, a French philosopher, August Comte, included in a list of discoveries he thought we would never be able to make, the chemical composition of the sun. Which just goes to show that we should never say, 'That's impossible!'"

What makes the sun hot?

"I made a point of arriving tonight before sunset. You told me you wanted to ask me some questions about the sun. Let's take advantage of its presence before it disappears."

"I'd like to know how long it's been there like that in the sky, and how it's able to produce so much light and heat."

"Those are questions we have been asking for thousands of years. But we've only known the answers for the last century. That's very recent.

First I'm going to give you the answers, and then I'll tell you how we found them. It's always interesting to look into the past and to see how problems were originally expressed, before the solutions were arrived at.

The sun's heat is generated by nuclear energy, just as in the reactors that are an important source of electrical energy in many countries. It has been shining for more than 4.5 billion years. The story of this discovery began with eighteenth and nineteenth century geological exploration. In excavating the underground strata, researchers found the fossilized remains of plants and animals that were alive hundreds of millions of years earlier. As life requires a permanent source of heat, that proved that the sun was already shining in this distant time. Which obviously raised a number of questions: what could the energy source be that has enabled our star to generate so much heat for so long a time? How has it managed to shine without exhausting its energy reserves? In the nineteenth century, scientists did not know about the existence of nuclear energy. It was discovered at the beginning of the twentieth century.

Suppose the sun were an immense ball of coal of the same volume, slowly being consumed. At the speed with which it burned and transformed itself into light to heat our planet, how long would it take for it to exhaust its reserves of energy ? The answer is simple: not more than one or two million years! There we have a problem: that would not be sufficient when we know that the dinosaurs already existed two or three hundred million years ago! And so it was presumed, logically, that there existed another form of energy, unknown at that time, that permitted the sun to shine for a much longer period. This form of energy, discovered at the beginning of the twentieth century, is nuclear energy. The sun, like almost all the stars, is made up primarily of hydrogen. At its centre, the temperature is fourteen million degrees Celsius. At this temperature hydrogen undergoes nuclear reactions that generate energy. These reactions transform the hydrogen into helium. Just like on Earth, in a hydrogen bomb invented by humans."

"But the sun doesn't explode!"
"That makes all the difference. In the sun,

energy is continually being generated. This is called controlled fusion. On Earth, we know how to make bombs, but we do not yet know how to control the output of energy. This is a very active research project.

This burning up of hydrogen at the sun's core is also the energy source for all the stars in our sky. It has two important consequences. First, the energy it gives off is transformed into light and heat: the sun's energy reserves are large enough that it can shine for ten billion years, which means we have no more problems with the age of the dinosaurs! Then, it produces new atoms. Four hydrogen atoms combine to make one helium atom. Later, helium itself is transformed into carbon, nitrogen, and oxygen. And later still, again through nuclear reactions, almost all the atoms in the cosmos are formed inside aging stars."

"But if atoms form inside stars, how do they get to us?"

"Stars do not live forever. They die when they have exhausted their reserves of nuclear energy. Our sun, according to our calculations, will die in about five billion years. It will then look like a huge nebula (wrongfully referred to as a 'planetary nebula')."

"Can we see any of those tonight?"

"Look at the sky. Do you see the Summer Triangle: Vega in Lyra, Deneb in Cygnus, and Altair in Aquila? There is a very beautiful nebula near the star Vega. But you need a telescope to see it. All the matter in a dying star including the new atoms formed during its lifetime are scattered into space, and later, are gathered into the nebulosity of the Milky Way. There, new stars may be born. Some will have planets like our Earth. And there you will find atoms born inside the dead stars."

"Can we know how the sun was born? And how it's going to die?"

"To learn about that, we're going to take a walk in the nearby woods. It's an oak grove. Oak trees can live for a very long time. Much longer than us. Sometimes a thousand years and more. No one can witness all the events in the life of a single oak tree. But walking in the woods, we can see oaks of all ages: baby oaks still attached to the acorns from which they were born, small oaks with only a few leaves, big majestic oaks, oaks that are already dying, and finally, on the ground, dead trees covered in ivy and mushrooms, slowly rotting away. In this way we can reconstitute all the stages in the life of an oak tree without having to wait for centuries.

Today, we are very familiar with the lives of stars. We know that they are born in certain regions of the galaxy that we call star nurseries. They form when the great gas nebulae collapse under their own weight. The one that gave birth to our sun and the solar system with its planets, meteorites, comets, and so on, we call the 'protosolar nebula.' We know how stars live and die. For each one, we can determine its age and the time it has left to live.

The sky over our heads is like an immense forest of stars. As with the oak trees, we see them at all different ages: the very young, stars at the halfway point in their lives (such as our sun), aging stars, and what remains of dead stars. We have before our eyes all the stages of our sun's life, its past and its future, and for this we do not have to wait the five billion years for it to live out its days."

How can we tell the sun's age?

"Now that the sun has disappeared and the stars have begun to shine, we can talk about them if you like."

"But you haven't told me how we can measure the sun's age!"

"Again, we have to talk about atoms. At the beginning of the twentieth century, thanks to the work of Pierre and Marie Curie, we learned that some of the biggest atoms, uranium for example, have a strange property. They are not stable. After a certain time, they break down and give off heat. We say they are 'decaying.' One kind of uranium, uranium-235, breaks down on average after a billion years."

"Do all these atoms break down at the same time?"

"No, it's progressive. That means that after a billion years, half the uranium atoms there at the beginning will have decayed. After two billion years, only a quarter will remain. After three billion years, an eighth will be left, and so on. We say that uranium has a 'half-life' of one billion years."

"Where do we find it?"

"We find it in small quantities in certain stones.

When you place your hand on them, you feel the heat. By concentrating these atoms, we create the fuel for nuclear reactors. And atomic bombs as well.

But these atoms have another use; they can measure time. We can determine the quantity of radioactive atoms in a stone: the fewer radioactive atoms, the older the stone. In that way we can calculate the age of stones on Earth, and also of meteorites."

"What's a meteorite?"

"It's a small piece of rock revolving around the sun, like planets do. They come in all sizes. The smallest are like pieces of gravel. When they enter the atmosphere, meteorites vaporize and leave a trail in the sky. Remember the beautiful shooting stars we saw last year in August! Some larger meteorites fall to the ground. They generally contain, in small quantities, different sorts of radioactive elements, each with its own half-life.

We were at first astonished by the fact that almost all meteorites are the same age. *Four billion five hundred million years.*

When the cosmonauts went to the moon, they picked up stones off the ground. They brought them back, and measured their age in the same way. Result: they were the same age as the meteorites!"

"Why are all these objects the same age?"

"Remember that stars and their planets are created together out of a nebula of gas and fine dust (*see page 24*). And so we conclude that the measured age of the meteorites and the lunar stones is exactly the same age as the protosolar nebula and therefore of the sun itself. This whole little world was born at the same time, 4.5 billion years ago!"

We are stardust

"Now the sun has set, and the sky is magnificent tonight. We see stars everywhere. In one of your books, you wrote that we are all 'stardust'. What do you mean by that?"

"That is one of the great discoveries of modern science. A discovery that links us to the world of the stars.

Touch your forehead while looking at the sky. Would you believe that the atoms that make up your body were born inside stars? That's exactly what astronomers have discovered, thanks to their telescopes and their patient work. As I've already said, it's very hot at the centres of stars millions of degrees Celsius and there nuclear reactions take place. New atoms are created, that accumulate within the star. Later, after the death and dismemberment of each star, these atoms wander through space. A certain number become part of the material that makes up our planet. They circulate in the solid ground and in the oceans. And one day, they make their way into the life cycles of all species, becoming the constituent elements for every individual. And more are introduced through your food intake all the time. We can truly say that we are stardust! In this sense, the stars are the great grandmothers of all human beings of all ages, and of all the living things on earth. After death, our body's atoms return to the cemeteries' earth. They can be reused in the composition of other living beings, plants or animals. Atoms do not die. They are continually recycled in a vast circuit that encompasses the entire planet."

"Will that go on for a long time?"

" Until the death of the sun in about five billion years. Then, our sun will go from yellow to red and will swell up enormously. It will become a giant red star, like the beautiful Antares Scorpio's eye in the Zodiac perfectly visible in the south in summer, just above the horizon. Our planet will become much hotter. Water will evaporate and the Earth's surface turn to desert. Later still, the stones themselves will vaporize. All our planet's atoms will return to space and become part of new nebulae. Perhaps they will form other planets that will one day be home to other little girls who will ask their grandfathers questions. And the recycling will resume up there as it continues here today.

People often ask me: 'Of what use are astronomy and telescopes?' Here we have an answer. Thanks to them, we have learned that the stars, distant as they may be, are in no way foreign to us. They have played an important role in our lives. Without them, there would be no atoms, and so no brains to formulate questions! It is worthwhile making the effort to understand what is going on in the universe, and how we have come to exist! In talking to us about the universe, science is talking to us about ourselves. It is trying to understand all the events that have taken place in the heavens and on Earth, and that have led to our own existence. It is telling us our own story.

Hives and galaxies

"I still have lots of questions to ask you."

"Let's go back to our deck chairs and resume our conversation. Now night has truly fallen, and we can see many stars in the sky."

"Yes, everywhere. There's no part of the sky where you can't see them. Are there that many everywhere in the universe?"

"No. There are immense spaces without any stars. We can't see that with the naked eye, but our telescopes can. In the universe, stars are grouped into large clusters that we call galaxies. Each galaxy contains about a hundred billion stars. If you see many stars, it means you are inside a galaxy. If we were to leave our galaxy behind, we would see very few. We can compare the stars in a galaxy to bees in a hive. Each bee in a hive is born there, lives there, and dies there. There are many hives, and each bee belongs to one of them. It represents, in a way, its family. In the same way, each star belongs to a galaxy. Our sun is a star in the Milky Way."

"Can we see our galaxy in the sky?"

"Take a good look at the sky. We see a pale white band that rises above the horizon, passes over our heads, and dips down towards the horizon in the south. That is our galaxy. We can only see a part of it: the rest passes under the Earth and returns towards the north. Because we are inside it, we cannot get an overall view of it. It's a bit like someone who, perched on the branch of a tree, can't see the whole tree, but sees the branches spread out around him."

"Can we see other galaxies?"

"Galaxies, even those nearest to us, are not visible to the naked eye. Except for three, that we can just barely make out on an especially dark night. In the autumn sky in the northern hemisphere, we can see the Andromeda Galaxy near the constellation of Cassiopeia (shaped like a W). We have to look at it with binoculars. Remember, when you see this white, oval spot, that its light left the galaxy almost three million years ago, when the ancestors of the first human beings were just beginning to walk upon the earth. In the southern hemisphere, we can see two more galaxies: the Clouds of Magellan. They are the closest to us. The others are much farther away, some of them thousands of times farther."

"How many are there?"

"With our most powerful telescopes, we can count more than a hundred billion. The universe we see is like a vast archipelago of galaxies in a gigantic ocean we call intergalactic space."

"With these telescopes, could we see all the

galaxies in the universe?"

"No. Even the most powerful telescopes cannot see the entire universe. Our observations are limited by a horizon beyond which we can see nothing. A bit like when you look out from the edge of the sea."

"What do you think is beyond this horizon?"
"Without a doubt, more galaxies."

"How many?"
"We have no idea."

"Might there be an infinite number?"
"Yes, it's possible. What is fascinating in studying the universe, is that anything can happen. Even things you could never imagine."

An expanding universe

"I read somewhere that the universe is expanding. What does that mean? That it's getting bigger? And if it is expanding, it's expanding into what? You'll have to explain it to me."

"When we want to discuss a question, it's always interesting to go back to its origins. Where does this idea of an expanding universe come from? Around 1920, large telescopes began operating in California. The American astronomer Edwin Hubble undertook to measure the distance and the movements of a certain number of galaxies."

"People really thought that they were moving around?"

"No one knew. Anything was possible! The results were so astonishing, so unexpected at first, that Hubble himself doubted their validity. He thought he'd made a mistake somewhere! It was his own students who were able to reassure him. He didn't know that he had made a discovery that was going to have a profound impact on the way we view the world."

"What did he discover that was so extraordinary?"

"He discovered that galaxies are not motionless in space. They move in respect to one another, growing farther apart. This in itself was not that revolutionary. But what followed would create an enormous stir. Galaxies, together, move in a very special way. The farther away they are, the faster they distance themselves from each other!"

"How can we picture that? I think you're going to start talking to me about your famous raisin pudding!"

"Yes, that comparison will help us to describe the phenomenon. In a batter containing yeast, we mix in some raisins. Everything goes in the oven, and we observe what happens. The batter, as it rises, carries the raisins along with it, and slowly they move farther apart. Now imagine that we are sitting on one of the raisins, and looking around us. We would see our neighbour raisins moving in a very particular way. The nearest to us move slowly, but those farther away move much faster. And yet everything is moving away in one shared motion. It's as though the pudding is expanding."

"Is it the same with galaxies?"

"Yes, that's the way it is for all the galaxies in the sky: the universe is in a state of expansion. That simply means that there is an overall distancing of one galaxy vis-à-vis the others. Which suggests that they were all closer together in the past, and that they will be increasingly separated in the future."

"Does that mean that the universe is getting bigger?"

"We have to be very wary of comparisons. They have their limits. The universe is like a pudding in terms of its movements, but it is different in form. The pudding has a centre and an edge. It extends into the empty space of the oven. Our universe has no centre and no edge. As best we know, there is no empty space around it. The universe is galaxies, everywhere. And they are all moving away from one another."

"I can't imagine that!"

"Don't be surprised. When we're dealing with dimensions so far removed from our day-to-day perceptions, our imagination has its problems. Biological evolution has accustomed us to much more modest dimensions. This is the price we pay for trying to explore the universe! But astronomers who work on these questions have learned to adapt their imaginations to these immense spaces.

All you have to remember is that everywhere in the universe there are galaxies, and that the distances between them are constantly growing. That is the meaning of the expression: 'The universe is expanding.' That and only that!

Let us pause briefly to consider this discovery. It tells us something important about our world: that it changes with time. It was different in the past, and it will be different in the future."

"What did people think before this discovery?"

"People generally accepted the view of the universe propagated by Aristotle some two thousand years earlier. According to the Greek philosopher, it always existed, and would always exist. Without change. For him, the universe was static, and its nature was fixed for all eternity. Of course, Aristotle acknowledged that there are things that alter: wood rots, metal rusts, mountains erode, and valleys fill in. But, he added, these are only events on our small scale. Trivial, in a way. On a grand scale, that of the sky and the stars, nothing, he insisted, ever changed."

"How did he come to that conclusion?"

"Aristotle was familiar with the work of the Babylonian astronomers before his time (they had for ages kept detailed records of what went on in the sky). The constellations always returned at the same time of the year. Nothing seemed to contradict the idea that the universe was unchanging. Aristotle felt authorized to declare that the cosmos was eternal. No beginning. And no end."

"But if I remember correctly, in those days astronomers had no telescopes."

"Of course, that's what makes all the difference. The modest artisans who put together the first optical instruments in Holland in the sixteenth century could never have imagined the impact their instruments would have on human thought. Nor

Galileo, when, with his telescope, he observed Jupiter's satellites in 1610, and discovered that the Earth was not the centre of the universe. Hubble's measurements were able to demonstrate that the universe is changing radically. It was denser in the past, and will be less so in the future."

"Does that mean that in the distant past it was much smaller? As small as a little dot?"

"No, not necessarily. It could have been big. Perhaps even infinite. It's not easy to picture. We'll come back to this later."

A history of the universe

"You've explained to me what the "expanding universe" means. You've insisted on the fact that the universe is always changing. How does that concern me? What difference does it make to me that these galaxies   that I cannot even see with the naked eye are moving away from each other?"

"We have a long way to go before we can give a good answer to that question. A universe where nothing would ever have changed for all eternity, as Aristotle claimed, would be a universe with no history. The discovery that the galaxies were all in movement drawing away from each other over time   made it possible for us to affirm that the universe has a history. We are now going to enter a new phase of our exploration: we are going to try to reconstitute that history. What, after all, is a history? It's the account of a series of events that took place in the past. That presumes that there were things that happened at certain times, such as the French Revolution for the history of France, or the Battle of the Plains of Abraham for Quebec. These episodes influenced what was going to happen later. Without knowledge of this past, we cannot understand the present."

"And so astrophysicists are like historians?"
"To describe the situation properly, we're

going to compare the astrophysicists' task, rather, to that of the pre-historian who is trying to learn about humanity's past. He investigates how our ancestors lived. Where did they live? How were they able to feed themselves and keep themselves warm? To answer these questions researchers create what we call 'excavations.' They go to places where there are traces of ancient dwellings. They collect ashes from the fires, primitive tools carved from flint, sculpted reindeer antlers. All of that helps them to reconstitute, with a little imagination, and quite authoritatively, our distant ancestors' way of life."

"Yes, I remember, you took us to visit the Tautavel Grotto near Perpignan. In the museum, we saw reconstitutions of our ancestors' lives, several hundred thousand years ago."

"We know a lot about the way human beings have lived since the time of Tautavel. But the farther we go back in time, the more fragmentary our information is. We are always discovering new dwelling sites, and a few skulls, some better preserved than others. And yet there are many unanswered questions.

What is important, when you want to describe a period out of the past, is that you have access to the fossils from that time. Otherwise, you can say nothing credible. I insist on that point. It's going to be useful for us in our history of the cosmos. And it's as true for human prehistory as it is for the history of the

46

cosmos."

"What fossils could there be for astronomers? There are no reindeer antlers in the sky."

"Of course, we're not talking here about arrowheads or painted grottos. There will be, on the other hand, radiation emitted at certain periods in the life of the universe. Or the kinds of atoms created during certain cosmic events. All have left traces that we can still identify today."

"I suppose that, like prehistoric fossils, these vestiges of the past will play the role of 'supporting evidence' for the plausibility of that history."

"You understand perfectly. But before going on with this story, I want to talk about the work of a certain Albert Einstein."

"The man sticking out his tongue in a picture?"

"Yes, that's him. He played an important role in all of physics. Where astronomy is concerned, he obtained a particularly significant result. We can demonstrate, through his theory of relativity (developed in 1917), that if the universe is expanding, then it must also be cooling down."

"Like a gigantic refrigerator?"

"Exactly. It's like what happens in a refrigerator's motor. When we compress a gas, it heats

47

up; when we let it expand, it cools off. We could say that the universe behaves like an enormous gas where the galaxies are its particles. Hubble's observations show that this gas is expanding, and so it is cooling off. This is a second element in our history. The first: the universe is expanding. The second: it is cooling off."

"Everywhere? Is the whole universe cooling off?"

"Yes, everywhere at once over the entire expanse of cosmic space.

And now I'm going to introduce you to another important figure in our story: the Belgian canon Georges Lemaître. About 1930, he had the idea of bringing together Hubble's observations and Einstein's theories. (Even earlier, the Russian astrophysicist Alexandre Friedmann had postulated an expanding universe based on Einstein's work.) He developed a scenario for the universe's past. Beginning with what he called 'a primitive atom' that was extremely hot and dense, the universe progressively thinned out and cooled. This is the first version of what would later become the Big Bang theory. At that time, the scenario did not have very much success in the scientific world. Few researchers were ready to accept it. When I was a student in the United States, it was hardly spoken of in the Physics Department. It made people uncomfortable."

"Why?"

48

"The idea of an initial explosion seemed inadequate and was not seen as a serious proposition. Many scientists were persuaded that the universe had no history. Everything changed thanks to a Russian astrophysicist, George Gamow, whom I was lucky enough to have as a professor. He was a kind of comic giant, who loved telling jokes during class. He was not afraid of Lemaître's scenario. The idea of a cosmos with a history didn't bother him. He added, most pertinently: 'We still have to have a way of testing it scientifically, to have proof.'"

"Always those fossils that we have to find!"

"Exactly! But where to look for them? He had the brilliant idea of using a well known property of matter. The hotter a substance, the more light it emits. In a blacksmith's workshop, iron glows in the darkness. First, it's red. If we raise the temperature, it goes yellow, then blue. It mimics the rainbow's hues, and becomes brighter and brighter."

"Is that true for any substance?"

"Yes, without exception. Even strawberry jam, if we heat it enough. On the other hand, a body that cools also changes colour, and becomes less and less luminous. It gets darker.

Let us suppose, Gamow said, that the Big Bang scenario actually unfolded the way Lemaître describes it. Let's take it seriously, so we can test it. It implies that, in the past, the universe was brighter. The farther back in time we go, the more cosmic matter becomes hot and luminous. If we go far enough back, we must get to a point where the quantity of light is prodigious, an ultra-dazzling 'flash.' The entire universe is light."

"But what happened to this bright flash?"

"That is the question Gamow was asking in 1948. Did this light completely disappear from the cosmos during the cooling off period? Or did there remain a trace of it that we could still observe today? A kind of fossil of those glorious moments? That would confirm that this scenario was an accurate description of the cosmos's first instants."

"If I follow you, the detection of this radiation would confirm the Big Bang scenario. Is that what happened?"

"Gamow had made some calculations. He concluded that such a remnant should exist today as radio waves, invisible to our eyes, but that we could detect with a radio telescope. They were discovered in 1965, almost twenty years after Gamow's prediction, and entirely by chance. This was a great moment for science, and, in fact, for all of human thought. We now had a confirmation of the Big Bang scenario: that the universe had a history and that this history was one of cooling, following on very high temperatures, density, and luminosity.

If there is a lesson to be learned from this story, it is that an unpopular idea can be correct."

"I suppose, on the other hand, that a popular idea can be false?"

"Yes. The universe is what it is. It has nothing to do with our opinions. The scientific community reacted correctly. Now, most astrophysicists have adopted the Big Bang theory. They take it seriously, and use it to explore the first moments of the cosmos."

"They consider it to be true."

"Here, we must be careful. Science does not say: 'This is the way it is!' It says: 'This is the way it seems to be' or even better: 'There is probably some truth here.' Still, there are many difficult questions, unresolved problems, points to clarify. The Big Bang scenario remains, for the time being, the best account of the cosmos's past."

"Are there other 'fossils' from the early universe?"

"There are several. Here's one: 'ashes' from the Big Bang are still with us. These are the atoms of hydrogen and helium."

"What do these atoms tell us?"

"They take us back to the time when the universe was one minute old. Its temperature was then a billion degrees Celsius. As in the sun today, nuclear reactions were taking place in cosmic space. They transformed part of the initial hydrogen into helium *(see page 22)*. The Big Bang theory predicts that only 10% of hydrogen will be transformed into helium, and 90% will remain intact. These hydrogen and helium atoms can be found today in stars and nebulae. Their respective quantities correspond to what the theory predicts. These atoms, remnants of the Big Bang, are fossils out of the past, just like fossil radiation. They are what remains of the great primordial conflagration. This agreement between the observations and the predictions of the Big Bang theory is one good reason to take it seriously. Even if, I repeat, the reconstitution of our history still presents serious problems. There is a lot of space, still, for many observations and theories. We must continue to be cautious. This must be the watchword for any scientist."

How old is the universe?

"You've explained to me how to tell the age of the sun. Can we also calculate the age of the universe?"

"There are several approaches to that. The first method consists in using the measurements made by Hubble, which showed us that the universe is expanding. With our computers, we can construct a digital simulation of the universe, a kind of scenario that traces its transformation during the expansion. Then we run the film in reverse: galaxies progressively come together. We continue until the moment when they join and are superimposed on one another. The reading on our meter is then 13.7 billion years. That is what we call the age of the universe.

A second method derives from the idea that, logically, the universe must be older than its oldest inhabitants. If not, then something is awry and we have to go back to the drawing board. Let us first consider the stars. In many cases, we can estimate their age. For example, the three stars in the Belt of Orion are about ten million years old. The beautiful blue stars of the Pleiades, visible in winter near the Milky Way (you must look at them with binoculars!) are about eighty million years old. Our sun is 4.5 billion years old. A globular cluster (a large group of stars) situated in the constellation of Hercules, is thirteen billion years old. And so we have dated many stars. Now, what is most important is that we have never found a single star whose age is greater than fourteen billion years.

A third method: we use radioactive atoms such as uranium and thorium. There are many with different half-lives. In Chapter 5 *(see page 27)*, they helped us measure the age of the sun. We can also determine the age of these atoms themselves, in other words the time that has elapsed since their creation within stars. These measurements are less precise, but here again they agree with earlier estimates. We have never found atoms that are clearly older than fourteen billion years.

And so we have three estimates obtained by three different methods. For galaxies and stars, the results are determined through the use of telescopes in astronomical observatories. For atoms, we use counters in nuclear physics laboratories that measure radioactivity. The results are similar. This agreement is very significant. If stars or atoms that were older existed in the universe, we would be able to detect them. Up to now, they have not been found. This lends credence to the Big Bang scenario."

"I have a problem. I'll try to explain it to you. When I was born, fourteen years ago, I arrived in a world that already existed. My parents were there. And you told me that when the sun was born, there were already other stars. But what was there before the Big Bang?"

"To answer you, I'm going to go back to the role of fossils in historical research. We can only talk believably about a period in the past if there are fossils to support our assertions. Otherwise there is nothing we can say. This is a general principle that applies everywhere.

What we call 'the age of the universe' is simply the moment prior to which we have no more fossils. Many researchers have proposed scenarios for before the Big Bang. But they have provided no justification, no proof. They remain pure speculation. Perhaps, later, new observations will enable us to travel farther into the past. It remains to be seen.

This does not mean that nothing happened before those 13.7 billion years, only that we know nothing about them. The distinction is important. I like to say that the Big Bang marks the horizon for our knowledge of the past. It's not a beginning, it's a horizon, one imposed by the limits to our observations and our theories of physics."

"Another question. We often refer to the Big Bang as a gigantic explosion throwing out incandescent matter to a great distance. Where did this explosion take place? Isn't that the centre of the universe? But you told me there was no centre. I don't understand."

"Once again, we must be careful with our comparisons. The image of an explosion must not be taken too literally. It implies the existence of two different spaces. A first space, full of explosive matter, dynamite for example, which will produce the detonation. And around this first space, a second great empty space where the ejected matter will spread out. That is a good description of explosions on earth, and elsewhere, but it does not at all apply to the universe. The difference is that the universe is just one space! Today it is full of galaxies moving away from each other. In the beginning it was an incandescent magma expanding everywhere at the same time."

"So the picture of an explosion is not worth anything?"

"We can only retain it if we keep in mind that the explosion did not originate in a given point in space, but that each point in this gigantic space exploded at the same time."

"How can we picture that?"

"There are always problems when we try to conceive of dimensions beyond our everyday reality. But in the end we manage to do so. We must always beware of visual representations based on the ordinary things around us."

Are we alone in the universe?

"Look up into the starry sky and see this blinking light. It's a passenger plane. Try to imagine what is going on inside. It's mealtime. A hostess is pushing a cart down an aisle. She is passing out trays to travellers who are unwrapping their cutlery. All this we can imagine, simply by following that bright dot. It is, however, very far away. Seen in our sky, it appears now to be leaving the Great Bear constellation, and moving into the Coma Berenices. Now we're going to play another game with our imaginations, not with a plane this time, but with a star in the sky. The North Star, for example: straight north, always true to its place in the heavens. Seen from here it's just a bright point, like the plane. We have no idea what goes on there. This time, you'll have to open wide the floodgates of your imagination! We might think that, like our sun, it has planets. Let's go closer. Let's imagine a place where, sitting on deck chairs, there is a grandfather showing his granddaughter a star in the sky. As it happens, it is our sun that they are observing. A tiny bright point seen from afar. The grandfather says: 'Imagine, near this star, a planet called Earth on which a grandfather is showing the sky to his granddaughter.'"

"I'm looking at the North Star, and it's fun to

picture all that. It's like a game. Do you think it's possible?"

"That's the big question. Are there people out there in the sky, perhaps different from us, but who, like we do, look up at the stars? Or is ours the only planet in the cosmos that supports life?"

"What do you think, grandpa?"

"I have no idea! It's a question that we humans have been asking for a long time. Up to now, we have no proof that life exists elsewhere than on earth! But be careful: that does not mean that it isn't there. It's simply a confession of ignorance. But as the proverb says: 'The absence of proof is not the proof of absence.' We just don't know. Perhaps there is life around billions of stars. It could also be that there is no life anywhere but on Earth."

"How might we know?"

"First, we have to define what sort of life we're talking about. Ants, as far as we know, do not lie out on deck chairs asking if they're alone in the universe."

"But they're alive!"

"What do we call life? Here on Earth, we know innumerable sorts of life, from bacteria to giant trees, and including cats, kangaroos, and so on. What do they all have in common? They are born, they live, and they die. They nourish themselves. They have offspring, and many other activities.

For the purposes of our discussion this evening we're first going to talk about a very particular case. We will ask ourselves whether there are beings who, like us, might have television sets available to them, and who would watch the news every night.

On Earth, we have been emitting radio waves for a little over a century. From our antennas, these waves fan out into space at the speed of light. In one century, then, they have travelled one light-century, or a million billion kilometres. In this gigantic expanse of space that has already been touched by our radio waves, we count several thousand stars. Many have planetary systems. Antennas on those planets could capture our signals."

"I see them in front of their sets."

"Imagine, on a planet thirty light-years away, listeners waiting impatiently for the latest episode of *Dallas*, an American series broadcast thirty years ago..."

"But if they can receive our programmes, we could also listen to theirs!"

"On Earth, radio astronomers have been listening for more than fifty years. Using powerful radio telescopes, especially the one in Puerto Rico in the West Indies (an instrument more than a hundred meters in diameter), they are trying to detect messages from extraterrestrial civilizations."

"But could they understand their language?"

"No, of course not. But it should be relatively easy to make out a certain structuring in the sounds. We could easily tell the difference between spoken languages and just incoherent noise, 'static,' like what comes out of a badly tuned radio."

"Have we already received anything?"

"Around 1967, there was a moment of great excitement. Some perfectly regular *bip... bip... bips...* were coming from a particular direction in the sky. Impossible to confuse them with static."

"And what was it?"

"In fact, they were waves emitted by a star rapidly rotating around itself. Its thin ray of light was sweeping the Earth at regular intervals. Like from a lighthouse by the sea. As interesting as this discovery was, it had nothing to do with research into intelligent life. It was a star called a 'pulsar,' of which there are many in the sky. Disappointment!"

"And that's all?"

"That's all. That is where we stand. We've received no coherent message that might suggest that we are dealing with an extraterrestrial intelligence."

"Perhaps they use different waves than we do. Technical inventions that we know nothing about. Who knows?"

"Yes. And we've tried to explore other

possibilities. In vain. Many listening programmes have been abandoned for lack of results. However, several groups of amateur astronomers continue this research. They have their data analyzed by the computers of volunteer Internet users seeking the proverbial needle in a cosmic haystack."

"Could we know if there are distant planets where life has appeared, even if no one is sending out radio waves?"

"We've recently seen an important development for this research, the discovery of planetary systems around stars other than the sun. They are called 'extrasolar' planets. We are now aware of several hundred. We assumed that such planets must exist. But now we have the proof."

"Among these planets, are there some that resemble our Earth?"

"For the moment, we have mainly discovered very large planets, the size of Jupiter and Saturn. Simply because they are easier to detect than the smaller ones like our Earth."

"Could these giant planets support life?"

"We don't think so. In any case, not life as we know it. But there is always the possibility that our idea of life is too narrow. That other forms of life, unknown to us, exist elsewhere. When the Europeans arrived in Australia, three centuries ago, they discovered new flora and fauna, different from what they knew: kangaroos, platypuses, and other strange beasts. You have to have an open mind."

"How could we know whether there are extrasolar planets where life has appeared, even if they are sending out no radio waves?"

"There is one possible piece of evidence. It comes from what we know about our solar system. Ours is the only planet to have an atmosphere of oxygen."

"Why us and not the other planets?"

"Because, precisely, we have life here! It manifested itself on our planet a little less than four billion years ago. At that time, the atmosphere was made up primarily of carbon dioxide. For more than three billion years, life existed only as microscopic cells like blue algae, in sheets of sea water. As they breathed, these organisms gradually transformed our atmosphere. Thanks to this phenomenon, oxygen made its appearance. If life were to disappear from the earth, its atmosphere would again become carbon dioxide, as on Mars or Venus."

"Does that mean that if we were to find an atmosphere of oxygen on an extrasolar planet, we could conclude that there was life there?

"Without being certain, that would be a very good indication."

"Why do you say that that would not be certain?"

"You know, in science, we learn to be cautious. There might be another possible explanation for the presence of oxygen on these planets. But all the same, that would be a fabulous discovery! For the first time, we would have good reason to believe that life exists elsewhere than on the Earth."

Nature is structured like writing

"Grandpa, you're telling me some really astonishing things. How did we learn all that? And how can I know if it's true or not?"

"You'd like to know if we should believe what scientists tell us? For that, I'll first have to tell you how science was born, and how it works.

For a very long time, human beings have been aware of phenomena in nature that were often quite strange, and they've been asking themselves questions about them. For example: What is thunder? Some thought that it was the voice of an angry god, and that they had to get down on their knees to ask its forgiveness. Why did the sun disappear during an eclipse? Was it true that a dragon ate it, and that they had to offer sacrifices so it would come back? Was spring water cool because there were nymphs that kept it so?

However, such answers didn't satisfy everybody. A little less than three thousand years ago, in ancient Greece, there were people seeking more convincing answers. It's then that they decided no longer to appeal to imaginary figures to explain their observations, but to seek in nature itself, and only there, the answers to their questions. And they found some interesting answers: eclipses were caused by the moon's passing in front of the sun; thunder was not the voice of a god, but a natural phenomenon that took place among clouds, and was later explained as the noise from an electrical discharge. There was nothing there that was 'supernatural.'"

"Why are these answers better? Why are they more believable than those that came before?"

"They are more believable if you can provide convincing proof. Because there are always doubts! Why should I believe this rather than that? Seeking *in nature* the answers to questions *about nature* is called the scientific method, and this method was enormously successful. And so over the centuries, we saw the emergence of physics, chemistry, biology, and also geology and astrophysics. Now several hundreds of thousands of people in many countries pursue this activity. Thanks to their work, every day we are probing further into nature's marvels. The inventors of this method deserve to have their names mentioned, in particular Anaximander, Anaxagoras, and Thales. These wise men lived in a small Greek village called Miletus, which is now in Turkey, on the shore of the Aegean Sea. We owe them a great deal.

Now imagine that one of them were to come back to us out of the past, and ask us about the results and the success of this method that they invented: 'What have you learned that we did not know in our time?' We would be tempted to take him on a visit to a great library of science, where he could see on the shelves, and read, millions of books and journals. But this approach would be tedious. Instead, let's try to summarize in a few sentences the new knowledge acquired after all that work."

"What would your answer be, grandpa?"
"I'll sum it up in two sentences. Here is the first: 'Nature is structured like writing.'"

"You'll have to explain..."
"Watch carefully. On a sheet of paper, I'm writing the letter G. I ask you: 'What is this?'"

"It's the letter G."
"Very good. Now I'm going to add an R. Than an E. Then another E."

"And then, grandpa? I don't understand. What are you getting at?"
"Wait! And now I'm adding an N."

"Oh! I understand now. That makes 'green', the colour green!"
"There you are. We had to put all these letters together for a picture to appear in your head: the colour green. We call that an 'emergent property'. The meaning of the word emerged when I put the letters in the right order. Written words, as you learned at school, are letters assembled in a precise order. A meaning is given to this grouping of letters. Dictionaries are put together to define that meaning.

That's the way it is, not only in English, but in many languages on Earth. And now we're going to play the same game with words. I'm writing on the blackboard the following four words: 'The poppies are red'. What did I get?"

"A sentence!"

"Yes, a sentence that itself has a meaning: it tells us what colour poppies are. We can now play the same game with sentences themselves. That will give us paragraphs. Then we put paragraphs together to get chapters, then chapters to get books, then books to fill libraries. All the libraries in the world contain all our knowledge."

"Yes, I learned that at school. But what is the message for our visitor?"

"I'm getting there. We're going to build an 'alphabet ladder'. On the bottom rung, there are letters. Above that there are words, then sentences, paragraphs, chapters, books, and libraries. As you can see, at each stage, the elements are constructed by combining those from the rung below to form the ones higher up. And each time we go up a rung, we have emergent properties."

"Do we know who invented that?"

"This method has existed for about five thousand years. It was born in the Middle East, in a region that is now Iraq and Iran. In the beginning, it was used primarily for accounting, and for religious and legal precepts. Later, it was imported into Egypt, into Greece, into the Roman Empire, then into all of Europe and America. At the same time, it spread to East Asia. Today it holds sway all over the world. All children learn it in school and use it to communicate among themselves, whether on the paper of books and newspapers, or on the Internet."

"*I understand. But I still don't see the meaning of your first sentence: 'Nature is structured like writing.'*"

"That's what we're going to look at now."

The rungs of nature

"So, grandpa, can you explain what you mean by 'nature is structured like writing?'"

"I'll give you some examples.

Let's begin with a substance that's familiar to us and very precious: the water from our taps. It's a molecule made up of one atom of oxygen and two atoms of hydrogen. Water has properties that are not at all like those of the atoms they're made from. We breathe oxygen in the air. We use hydrogen to fill our balloons. But water is different: we drink it. Water is like a word composed of those atoms that are like letters. This is a good example of the resemblance to writing. Simple elements are combined to obtain new substances with emergent properties.

Another example: nitrogen is a constituent of our atmosphere used in its liquid state by businesses involved in artificial cooling. Let us put together one nitrogen atom with three hydrogen atoms. We get ammonia, a substance rather disagreeable to breathe (it smells like cat pee), but that is very useful for disinfecting hospital rooms.

Yet another example: we combine two atoms of carbon with six atoms of hydrogen and one atom of oxygen. We obtain the alcohol that we drink: wine, beer, whisky, or vodka. The patriarch known for his ark, the famous Noah, experienced one of the emergent properties of alcohol: that of provoking alcoholic intoxication (Genesis 9:20-9:21)!

And I'll give you one more. Table salt is made up of two atoms: chlorine and sodium. Chlorine is a corrosive substance that we find, among other places, in bleach. Sodium is a metal. Combined, these two atoms form table salt (the molecule sodium chloride), which we add to our food to enhance its taste. All these discoveries came down to us from eighteenth and nineteenth century scientists, like Lavoisier, Priestley, and Dalton.

We're now going to build a ladder of nature analogous to the ladder of alphabets."

"I suppose that the atoms are going to make up the bottom rung like the letters of the alphabet, and the molecules will be the words."

"You're on the right track. That idea comes from the Greek philosophers of antiquity. In particular, Democritus and Lucretius. They represented the atoms as little unbreakable balls; the word 'atom' means 'unbreakable' in Greek. Through various combinations, those atoms formed, in their opinion, all of nature's substances. At the beginning of the twentieth century, physicists built particle accelerators, scalpels of a sort, that made it possible for them to study atoms. They then discovered that, far from being unbreakable, these are complex objects with an internal structure. At the centre there is a massive nucleus made up of protons and neutrons. Around it, there are orbiting electrons.

We owe this discovery to Ernest Rutherford, in particular."

"That makes me think of the solar system with the sun at the centre and the planets around it."

"There is a similarity, it's true, but there are also great differences. Remember, we must always be careful with our comparisons. These atoms represent, for nature, another opportunity to play with alphabets. Their nuclei are the words for which the protons are the letters. The nitrogen nucleus contains seven protons. That of oxygen, eight. Iron, twenty-six. Lead, eighty-two. For each number of protons there is an atom in nature. There are about a hundred. The lightest atom hydrogen   contains just a single proton. The next is helium, with two protons. These are the oldest atoms, the first to be formed in the universe, practically at the time of the Big Bang. They are what remain of the conflagration during the cosmos's first seconds. The other atoms, carbon, oxygen, iron, gold, etc., right up to the heaviest, uranium, which contains ninety-two protons, were formed within stars (*see page 23*).

"But the protons themselves, are they unbreakable?"

"I was waiting for you to ask that question. I'll answer you with a recollection from my student years. In his astrophysics classes, George Gamow taught us that the word 'proton' came from the Greek 'protos,' which means 'first.' Then he added, 'Now we have reached the Greek philosophers' bottom of the ladder. Atoms are breakable, but not protons. They are "the first." They have no internal structure.' In response to our questions, he said: 'Yes, I understand your skepticism, since we've been able to smash atoms, which were considered unbreakable. But now it's the end of the road: I'm prepared to bet half my fortune on it. We'll never smash protons.' Too cowed by this illustrious cosmologist, we made the error of not accepting his wager. Because Gamow came from a very rich family.

In fact, a few years later, some clever experiments proved that protons (and neutrons as well) were not simple particles. They were made up of three 'quarks.' This lower rung was discovered around 1970, and the credit goes in particular to Murray Gell-Mann.

There are six varieties of quarks in nature. Physicists have given them the names of letters, accompanied by imaginary words, just to amuse themselves! There is the 'u' (for *up*), the 'd' (for *down*), the 's' (for *strange*), the 'c' (for *charmed*), the 't' (for *top* or *truth*), the 'b' (for *bottom*).

The proton is made up of two u quarks and one d quark. The neutron is one u quark and two d quarks. All the possible combinations of these quarks, whether two by two or three by three (like words with two or three letters), make for a rich variety of particles whose existence we've been able to verify in large accelerators. Almost all these particles are unstable.

They break down and disappear in billionths of a second! The neutron is unstable. When it is not part of a nucleus, it disappears in twenty minutes. The proton is stable."

"I guess you know what I'm going to ask next. Is the quark breakable?"

"After the episode of supposedly 'unbreakable' atoms, and supposedly unbreakable 'firsts,' no one today would dare to assert that we have finally reached the lowest rung, that of the particles analogous to letters, which we call 'elementary particles.' To probe this question, we would need even more powerful scalpels. Recently, in Geneva, a great accelerator consisting of a tube twenty-seven kilometers in circumference, buried a hundred meters underground, was put into operation to probe even further the structure of matter. Perhaps we will learn more about the nature of quarks. For the moment, we can say no more, pending some proofs derived from appropriate observations."

"But you didn't talk to me about electrons."

"The situation is the same as for quarks. We just don't know. To build our ladder, we're going to assume for the time being that quarks and electrons are elementary particles. That they are on the bottom rung. So let's sum up. We've explored three rungs. The quarks, on the first rung, are the 'letters;' the protons and the neutrons, on the second rung, are the 'words' of atomic nuclei; and the atoms, on the third rung made up of these nuclei and these electrons constitute the 'sentences' of molecules."

"And what do we find on the higher rungs? Living beings, I suppose, made up of molecules?"

"Yes, we come to the rung of living cells. We can see many varieties with a simple microscope. For example, in a drop of water taken from a pot where flowers have been sitting for a long time, myriads of little organisms are moving about in all directions. Now, and this is of particular interest to us, biochemists have taught us that these little cells are combinations of molecules, like proteins and DNA. Each molecule plays a specific role within the cell, inscribing instructions for the genetic code, or participating in the cellular activity that creates different hormones."

"So we've just climbed another rung on our ladder?"

"Right. And now we're going to climb yet another rung where the cells themselves are constituent elements. These little organisms are going to combine to form plants and animals, including our own bodies. It's a kind of federation in which each cell puts its skills at the disposition of the whole. We owe this discovery to the German chemist Theodor Schwann, who wrote, about 1860, that 'the cell is the basic unit of the animal and vegetable kingdoms.' Some cells capture sunlight to provide the organism with energy. Others digest food. Still others produce offspring. Your body, like that of all animals and all plants, is made up of cells. Red blood corpuscles transport the oxygen that you breathe to the neurons in your brain, and make it possible for you to talk to me. In your eyes, cells take in light and send images to your brain. Thanks to the combined activity of these myriads of cells - you are alive!"

"Grandpa, are there other rungs above these? Combinations of living organisms?"

"Let's take the example of a beehive. Every bee is assigned a function: to go and gather pollen from flowers, to repel intruders, to control the temperature. Here the bees are the basic elements, and a harmoniously functioning hive is the result."

"Like ant hills and termites' nests."

"Yes, but we can also talk about an orchestra playing a Mozart symphony with different musicians and instruments: violins, violas, cellos, flutes, oboes, clarinets, bassoons, and so on. At a concert, the music you enjoy is the emergent property of the various performances by the musicians under the baton of the conductor.

Another example: exploring the moon. To prepare the rockets and the probes, to train the astronauts, hundreds of thousands of people combined their efforts with a specific goal in mind: to land on our satellite. No one on his own could accomplish this task.

That too is an example of emergent properties derived from the combination of diverse elements.

I'm now going to sum up the situation by again taking your own body as an example, the body you shower or plunge into a swimming pool, and that you touch with your hand. In the last analysis, it is made up of quarks and electrons. And there are a lot of them! About a hundred billion billion billion (100 000 000 000 000 000 000 000 000 000). One plus twenty-nine zeros. It varies according to one's weight, but not that much. Now close your eyes and say to yourself: 'I exist.' Open your eyes, and say 'The world exists around me.' You have just performed an astonishing feat, among the most extraordinary success stories in the universe. For you to become conscious of your own existence and that of the world around you, a hundred billion billion billion quarks and electrons, woven together in a structure of unimaginable complexity, have had to play their individual roles. Like in a clock where every gear has to function properly, your quarks and your electrons are all in place to make it possible for you to act: to read, to concentrate, and to sleep when you have to.

And that is the meaning of this first message for our extraterrestrial visitor. That is the meaning of the sentence, 'Nature is structured like writing.' At the same time, it sums up what we have been taught by:

1. Physics: the combining of quarks into protons and neutrons and of these into atomic nuclei, which combine with electrons to form atoms;

78

2. Chemistry: the combining of atoms into molecules;

3. Biochemistry: the combining of molecules into cells;

4. Biology: the combining of cells into living organisms.

Each of these sciences represents another phase, as matter is organized in the universe.

The second message, for our guest from Miletus, derives from astronomy. It says this: 'The ladder of complexity builds itself up over time.' We will come back to that during our next conversation."

Blaise Pascal and the top
of the ladder

"You've told me about the history of the universe since the Big Bang. You've talked about atoms, molecules, and so on. You still have to explain where life came from. That's my favourite part. I'd like you to talk about my cat, Coquecigrue. When does he come into the story?"

"Just be patient for a while longer! We've already climbed five rungs on our ladder. At the sixth rung, we find living cells: blue algae, pretty diatoms, and all the little beasties we find in the water of faded bouquets. We'd very much like to know when they made their appearance in the universe."

"It seems you have a problem there, because if I understood correctly, we don't know if there is life anywhere else but on Earth."

"Yes, you're right. We can only chart our history on our own planet. Where the rest of the cosmos is concerned, we know nothing."

"So let's go back to our solar system. You said that the sun and its planets were born 4.5 billion years ago. So life on Earth must be more recent. What do we know about that?"

"Geology tells us that at its birth the Earth was

an incandescent ball of lava of more than a thousand degrees Celsius. No life was possible. Water could only exist as a vapor. No organism could have survived under those conditions."

"You're talking, of course, about life as we know it."

"You're right to say that. However, we know that some hundreds of millions of years later, the Earth had cooled, the vapor condensed, and multitudes of tiny organisms were swimming around in liquid pools."

"What had happened in the meantime?"

"No one really knows. When Louis Pasteur, in the nineteenth century, proved that, contrary to what had been believed, life did not appear on its own, he left us with a big problem. If 'spontaneous generation' no longer takes place, how did it occur in the beginning? How were molecules swimming in water able to combine in such a way as to form an organism able to nourish itself and to reproduce? That remains one of the great mysteries of modern science."

"We have no idea?"

"We do, but there is to date no really satisfying scenario. What is important for us is that, whatever the series of episodes leading to that result, life did appear and we are the proof. We are all descendants of those tiny beings that proliferated in the ancient oceans.

Where the rest of the story is concerned, we are in familiar territory. Here, the basic elements are the molecules that make up the organisms, and the emergent property is, quite simply, life."

"Life! What an incredible event! Do we know when it appeared?"

"More than three billion years ago, and less than four. Even that is still uncertain. The oldest signs of life have been found in Australia and Greenland. They are microscopic organisms whose skeletons, which accumulated after their death, formed large rock-like structures. In geology, they are called stromatolites."

"You did say that by breathing, they produced the oxygen that we breathe now?"

"These little structures multiplied to the point where they were able to affect the planet itself. Acted upon by light, they were able to use carbon dioxide to produce oxygen. They enabled living beings to develop more efficiently. They opened the way to an accelerated evolution. Without them, we would not exist. But it took another three billion years before we reached the next rung."

"On Earth! It could have been different somewhere else!"

"Perhaps. We still don't know! A little less than a billion years ago, a new chapter of life on earth began. We saw the emergence of beings where several different kinds of cells were combined. In the course of a slow evolution, there appeared fish, amphibians (that would come out of the water, like frogs), reptiles, birds, and mammals, including monkeys, hominids, and... us."

"At last my dear cat Coquecigrue!"
"More exactly, his ancestors the felines... among other mammals."

"Are the basic elements now cells?"
"We can distinguish about two hundred different varieties, each with its own specialty. In humans, the neurons, or brain cells, play an extremely important role. They make it possible for us to think and to ask questions. Our intelligence is an emergent property when they are combined."

"Closing my eyes and becoming aware of my own existence, like you suggested?"
"This amazing feat has been possible only for a short time compared with the age of the universe. Some millions of years compared to fourteen billion."

"Why did it take so long?"
"It took a long time, first of all, for the stars that form the atoms to be born, to live, and to die. Then for solid planets to form, where water could collect. And finally, because biological evolution is

slow, to go from the amoeba to beings that could think. All that took billions of years!"

"Are there still more rungs on our ladder?"

"Remember the hive: every bee, with its own special task, contributes to the harmonious functioning of its environment. The same for the symphony orchestra, about which we have already talked *(see page 77)*: the different instruments are coordinated by signs from the conductor so as to produce a unique result, such as Beethoven's Ninth Symphony."

"Your second message for the ancient philosopher, 'The pyramid of nature builds itself up over the course of time,' leads me to ask a question. Will there be other stages in the future? You have given me dates out of the past for climbing the ladder. The last, that of organisms, was less than a billion years ago. That of animal and human societies is even more recent. It's hard to believe that the story is over."

"I love it when you ask that kind of question. I will add this: could the first cells, which appeared over three billion years ago when molecules combined at the bottoms of the oceans, have suspected (in a manner of speaking) that they themselves would be brought together into living organisms? Just like the organization of matter in the universe, the increase in complexity has been going on for fourteen billion years, and no one knows what the future has in store.

To shed light on the two messages we are

confiding to our gentleman from Miletus, we are going to summon a man who gave a lot of thought to these questions. He is Blaise Pascal, a seventeenth century philosopher. You know his famous declaration: 'The eternal silence of these infinite spaces frightens me.'"

"Yes, I learned that at school. I'm not sure I really understand what it means."

"Pascal wrote that sentence only decades after Galileo's discoveries. He now knew that the Earth, far from being the centre of the universe, as everyone was happy to believe up to that point, was a small planet lost in endless space. The discovery of the universe's immensity made him dizzy. He felt lost in these immensities that were totally foreign to him, utterly indifferent to his existence."

"Everything you've taught me about the universe, he didn't know. What could you tell him to reassure him?"

"We could tell Pascal that, if the universe was not as immense and as old as we now know it is, he would never have been able to write those words. He would never have been born if the universe had had the limited dimensions and the six thousand years of life proposed in the biblical texts. That is our message for Blaise Pascal!"

The stone tablets

"Tonight I have a big question for you. You've told me the story of the universe. I understand that during the cooling of the initial incandescent magma, more and more structures, increasingly complex, were formed. And that in fact, I owe to the stars the atoms and molecules in my body."

"Yes, we live in a universe where marvelous things happen. Not only has life flourished, but it has brought us the music of Mozart and the poetry of Verlaine."

"I can't help asking whether that doesn't mean that there must be a great architect who conceived all that, or a programme, like in a computer? And therefore a programmer?"

"That is a natural question for all of us. I will approach it by asking two other questions. What they have in common is that we have no answer for them."

"So why ask them?"

"Because the fact, on the one hand, of acknowledging their existence, and on the other hand of confessing our ignorance, are important steps in our thinking. The first question seems very naïve, but it isn't. We owe it to the philosopher Leibniz: 'Why is there something, rather than nothing?'"

"We could answer that if there were nothing, no one would ask that question!"

"Of course! However, it would appear that there is 'something.' That is a fact. But we have no answer for the question, 'why,' and so we have to confess our ignorance. And that leads to the second question."

"What is that?"

"Here it is: why did this 'something,' which revealed itself at the beginning of the cosmos as undifferentiated chaos, gradually structure itself instead of remaining chaotic? We ask that question in the light of our scientific knowledge, and in particular because of what we have learned from the Big Bang theory and its credible scenario for the history of the universe."

"Does science give us any answers?"

"In a way, yes. It tells us that it's thanks to the existence of what we call the 'forces' of nature, and the 'laws' that rule it, that structures have been able to emerge. There is the force of gravity for the planets and stars, the electromagnetic force for atoms and molecules, the nuclear forces (there are two) for protons and atomic nuclei. We know these forces well. Their properties have been accurately measured in our physics laboratories."

" Yes, I understand. But it seems to me that you

*are just putting off the question: if I ask you why there
are forces, rather than there not being forces in nature,
what would you say?"*

"You're absolutely right. Even if we found an
answer, one could repeat the question: 'But why this
answer?' and so on. The trail of 'whys' and 'becauses'
would be endless. Again, we're going to have to confess
our ignorance. We're going to say: there are forces, and
it is these forces that make the structuring of matter
possible. In short, we do not know why there is
'something' rather than nothing, and why there are
forces   rather than no forces   that have made it
possible for this 'something' to become structured, and,
in particular, to give birth to you, to me, to your
parents, to your cousins, and so on."

"You said that we know these forces well?"

"To introduce you to them, we're going to
have to make a detour via the history of science. Before
the work of Galileo and Newton, in the seventeenth
century, the generally accepted picture of the cosmos in
the scientific world was that of Aristotle. For him, the
universe was made up of two different parts: one below
(under the moon), and one above (over the moon).
Below is our earthly world, made up of corruptible
matter, subject to change: wood rots, metal rusts,
mountains erode, valleys fill in. Above, there are the
stars, the sun, the planets, made up of a substance that
is pure, incorruptible, inalterable, always true to itself."

"Why was the moon made the frontier between the two worlds?"

"For the Ancients, the moon had a dual status. It both changed (the different phases), and it did not change, because like the stars, it returned regularly at foreseeable dates. Then Galileo looked at the sky with his telescope. He discovered Jupiter's satellites, the phases of Venus, and the lunar mountains. He concluded that there were not two worlds, but only one. And it was a few decades later that the legendary tale of Isaac Newton took place. One moonlit night he saw an apple fall from a tree. That caused him to reflect deeply on what it meant. It moved him to demonstrate that it's the same force   gravity   that makes apples fall, and that makes the moon revolve around the Earth, and the planets around the sun. The same forces are at work on Earth and in the solar system. That was the birth of astrophysics."

"Is it true everywhere? Even for the most distant galaxies?"

"We found the answer to your question in the twentieth century, when the giant telescopes came into service. We collected the light emitted by oxygen atoms in stars that are billions of light years away. We compared the photons that have traveled for billions of years with photons emitted by an oxygen source in the laboratory. What we found is that the 'old' photons from the galaxy obeyed, with great precision, the same laws as the 'new' photons from the lamp. These experiments, along with many others, showed that the laws regulating the properties of forces in nature are the same everywhere, throughout space and throughout time. I should add that if it were otherwise, if the laws changed according to the place and time, it would be much more difficult to study the cosmos. We may regard mother nature's consistency as a gift to the poor scientists who are trying to understand her!"

"But it seems to me that there is a problem here. You showed me that we are living in a universe that is always changing, and now you're talking about laws that do not change."

"Yes, this paradox is very real. On the one hand the scenario for the Big Bang has been authenticated, and on the other the universality of these laws has also been well established. We have often asked ourselves why the magmatic matter of the primordial universe organized itself into structures at all scales of magnitude, and why at these scales it was progressively populated? The answer: because there are forces that act upon particles and ensure that they are structured. We have discovered that the laws governing these forces have a remarkable attribute: they are the same always and everywhere, even while everything is changing in the cosmos."

"That makes me think of the stone tablets of Moses, on which the Ten Commandments are engraved in the Bible story. On what 'tablets' are the

91

laws of nature inscribed, for them to be so permanent?
Another question with no answer."

"These laws never cease to astonish us. Here is
a more recent discovery that no one anticipated when
we began to study the cosmos. Our observations, and
our theoretical models of the universe, show that these
laws exhibit just the properties one would need for life
to appear."

"You mean that if they were different, life
would not be able to appear? How can you prove
that?"

"Computers have made an important
contribution here. We simulate through our
calculations what would happen in a universe subject
to different laws. We call these 'toy universes.' Every
universe is subject to a collection of laws that are, in a
sense, the recipe for its pudding. We posit, in the
beginning, an extremely hot magma, dense and
luminous as in the model for the Big Bang. We let it
cool while observing what is going to happen. We find
that in each, in a toy universe just as in the real
universe, there is a cooling, a thinning out, and a
darkening of cosmic matter. But here is what is
important: there are important differences, according
to the recipe chosen at the start.

In the great majority of cases, life as we know
it cannot develop. There are instances where no galaxy,
no star, no planet manages to condense out of the
initial stew. There are no solid planets where liquid

92

water can settle, for life to be born. In other cases, all of matter fragments and rapidly contracts. The result is a population of very dense stars that emit no light (black holes). No sun and no planetary system. Or else all the hydrogen is transformed into helium in the very first minutes (our third rung), and no water can form later on to foster the first living cells (our fifth rung). Or there is too little carbon, a crucial element for biochemistry. In many cases, there is no star with a long enough life span for life to appear and evolve in its planetary system."

"You're telling me that the laws of physics in our universe have just the right properties for a questioner to appear. What is certain is that if that were not the case, we would not be here talking to each other. There would be no one! In other words, we (our universe) have won the lottery! We've been very lucky!"

"The scientific community is very divided on that question. Some see it as banal and of no interest. Others, on the contrary, consider this information to be very important. As one might expect, philosophical and religious feelings play a role in those personal reactions."

"And you, grandpa, what do you think?"

"I can't help thinking that there is something very interesting here, but that it escapes us completely. We will come back to this when we talk about parallel universes."

The multiverse

"My friends, who know we've been having these conversations, are interested in what you call parallel universes. Are there other universes like ours, but completely separate?"

"There is a lot of talk about that possibility in scientific circles. According to some authors, our universe could be just one cosmos among myriads of others. All those universes together are commonly called the 'multiverse.'"

"What do you think?"

"Everything is possible, of course. To claim that there could be no other universe than our own does not to me seem very scientific. The problem is that for the moment, we have absolutely no proof, even indirect, for the existence of even one other universe. In science, we accept new ideas, but in exchange we insist on proofs, on confirmations from appropriate observations. Otherwise it's science fiction."

"You said the other day that 'absence of proof is not proof of absence.'"

"That's very true. It's why I have an open mind on this question. There is, however, an argument that has been adopted by a number of astrophysicists that would seem to support the idea of this 'multiverse.'

Let's go back to our own universe, and to the complex of laws that regulate it. Most specifically, the finding that these laws have exactly the right properties to make possible the emergence of life and of consciousness."

"Yes, but I don't see the connection with the multiverse."

"Now imagine that in this multiverse, each universe were subject to different laws (a different pudding recipe). The result? Those that do not have our recipe, which we will call the 'fruitful recipe,' will cool, will thin out, will darken just like ours, but will remain sterile. Either no long-lasting star will form, or hydrogen will all be turned into helium, and as a result no water molecule will be able to form."

"And then? What would happen in the end?"

"In this context, the researchers say: 'If we can ask questions, it's just because among all these universes, we are living in one that is fruitful.' And so for them, everything is very simply explained, and that justifies their belief in the multiverse."

"You did say that we have no way of knowing whether these universes actually exist."

"For the moment, no! Perhaps the situation will change, and we will be able to flush them out later on with new instruments. But for the time being, we have no way to verify either their existence or their non-existence."

"*Tell me if I've understood rightly. The believers in the multiverse say: 'If there exist other universes, and if these obey laws different from our own, we have all the more reason to be astonished at the fruitfulness of ours. Our universe is only one among a profusion of others, and is only set apart in that it has productive laws.'*"

"Yes, that's the idea."

"*It's a pretty one, but it seems that there are a lot of 'ifs.'*"

"I agree. Too many 'ifs' for my tastes. But if we cast doubt on the existence of the multiverse, how can we interpret the fact that the cosmic laws are exactly those that require the growth in complexity and the emergence of consciousness on Earth (and perhaps on other planets)? That is the situation where we find ourselves today. I have no answer.

It's important, I think, to keep an open mind. It's better to accept that there are some questions with no answers, than to adopt unsatisfactory solutions. Otherwise we risk closing doors that could open onto promising perspectives. The observations that led us to conclude that the laws of physics are well adapted to the emergence of life and consciousness, would not have been able to deliver their potentially interesting message. That would have been a shame."

The clock and the clockmaker

"You've talked to me a lot about the cosmos. You've described how matter was organized. You've shown me that the cosmological laws are exactly those that can produce this organization. So who made up these laws? Doesn't such a beautiful creation need an inventor?"

"This is an area where we must be very cautious. Here we move away from the domain of science, and enter the domain of factual interpretation. In science, we have proofs. Here, we can find none. We can only formulate personal opinions. To begin, I'm going to tell you a story.

In the distant past, long before the great explorers roamed the world and we had photos taken from satellites, people wondered about the shape of the Earth: 'Is it flat, or is it a ball?' Some said: 'It couldn't possibly be a ball, because then those people living on the other side of the ball would have their heads pointed downwards. They would fall into empty space.' This reasoning seemed perfectly logical, but it was false. The Earth is round and the Australians don't fall off! Where was the mistake? In the meaning of the word 'downwards.' 'Down,' as we now know, is always in the direction of the centre of the planet. But at that time, no one knew. Their ignorance skewed their conclusions.

This story teaches us something very important: we always reason on the basis of what we know. And it shows how dangerous it is for us to apply our reasoning to other dimensions. Our arguments are valid in the context of our science at a particular moment in time. And when new knowledge arrives, our thinking must adapt. Voltaire said: 'I cannot imagine that this clock exists without there being a clockmaker.' This reasoning is valid at the scale of clocks and clockmakers. But can we really say that the universe is like a clock? We must always be wary of comparisons. What is a clock? It's a mechanism made up of many gears. In Voltaire's time, around 1750, the structure of the solar system had just been discovered, with its planetary orbits. It's understandable that Voltaire would have compared it to a clock. He extended his argument to the entire universe, about which, at that time, little was known. Today, modern physics gives us a much more complex and mysterious picture. We are far from solving the mysteries of atomic physics. We don't really know what reality is."

"Does that mean we must abandon the idea of a great architect?"

"I don't know. It's a question I've asked myself for a long time. One thing is certain: Voltaire's answer is grossly inadequate. But what can we put in its place? I see your little cat Coquecigrue asleep on your lap. You told me once that he was very intelligent."

"Yes, he's amazing. I have the impression, sometimes, that he's thinking."

"But you'd never try teaching him geometry, for example."

"No, of course not, he wouldn't understand it!"

"I often feel that just like geometry is beyond the ken of your little cat, these questions concerning the nature of the universe are beyond our grasp. They outstrip our brainpower. Despite all the advances of modern science, the universe remains deeply mysterious. Perhaps that will continue indefinitely. I think we have to contemplate that being the case. But who knows?"

What is a black hole?

"I've heard a lot of talk about black holes. Do they really exist? Are there some in the sky over our heads? And besides, if they're really black, we shouldn't be able to see them!"

"The answer is yes, there are billions of black holes. As big as the solar system, as small as a mountain, and perhaps even smaller. In fact, 'hole' is not a good word for them. They are not holes, but rather strange stars. To tell you about them, I'm first going to indulge in some make-believe. Imagine that tonight an enormous genie approached the sun and began to squeeze it between its huge hands. Imagine that our star, which is about a million kilometers in diameter, found itself reduced to only three kilometers."

"What would happen?"

"Tomorrow, there would be no sunrise; it would be invisible!"

"Why?"

"Because it would have become so dense, so compact, that no light could escape from it. It would fall back on itself like water in fountains."

"What would prevent it from escaping?"

"The gravitational attraction of matter that has been so compressed! Just like the attraction of the Earth prevents the stones you throw from leaving our planet. A black hole is a star so dense and compact that nothing can escape from it. Not even light! Whatever falls on it never returns. It's a kind of giant vacuum."

"Could the Earth be swallowed up by it?"
"No, it's much too far away. It's beyond its range."

"But tomorrow morning, how could I know the sun was still there, since I couldn't see it?"
"By observing, night after night, the stars still perfectly visible in the sky, you would see the seasonal procession of constellations, just like before. That would show you that the Earth was still revolving about the sun."

"So even if it were to become a black hole, the sun would still attract the Earth and keep it in orbit around it."
"Exactly. Our sun acts upon its planets in two different ways: first, it shines light on them, and second, it attracts them through what we call its gravitational field. All bodies have this property. They attract each other mutually, and the more massive they are, the more they attract what is around them. However, the two activities are independent of each other. Even if it sent out no more light, the sun would

continue to attract the planets. A black hole makes its presence felt through its gravity.

Let's imagine another chapter in our story. This time the evil genie slightly increases the suns mass."

"What would happen to the Earth? If I understand correctly, it would be more strongly attracted by the sun. Would it fall into it?"

"Not necessarily. It could just be drawn closer. It would turn more quickly, at a shorter distance than now. And if the genie removed some of the sun's mass, the Earth would back off and would turn more slowly. This little story will be useful for us later on. It shows how a star can influence the movements of other stars around it, even if it sends them no light."

But let's go back to the black holes. We know there is one at the centre of our galaxy, the Milky Way. Just like the planets revolve around the sun, we have recently observed several stars that orbit around this invisible star. By measuring their speed, we can determine its mass, which is three million times greater than that of our sun. In our latitudes, in summer, this black hole is situated in the direction of the constellation Centaurus, near the southern horizon. It is close to Antares, the beautiful red star, Scorpio's eye."

"It's enormous! Couldn't it attract us and swallow us up?"

"No, we are safe because we are so far away. It

is now generally agreed that every galaxy has a black hole at its centre. The Andromeda galaxy has one, thirty times as massive as ours. In some galaxies they are much larger still, up to a thousand times the size of our own. These monsters swallow up entire stars and nebulae. Avalanches of matter rush into them. Before disappearing forever, these gaseous fragments in free fall send out flashes of light on all wave lengths: radio, infrared, visible, ultraviolet, X-rays, and gamma rays. These 'swan songs' are detectable all through the universe. We call them 'quasars.' We then say that the monster 'awakens' when he is 'fed.'

"Are there also smaller black holes?"

"Yes, they form when a giant star dies. After the explosion that marks its end, part of the star's matter falls back on itself and becomes as dense as if a great oil tanker were crammed into a thimble. There are billions in our Milky Way, as there probably are in other galaxies."

"Are there smaller ones still?"

"To date we have seen no sign of their existence. But there is doubtless more to this fascinating species than astronomy has yet been able to show us."

Dark matter

"My friends have been asking me if dark matter is made up of black holes. What is this dark matter that seems so mysterious?"

"To talk about that, I'm going to first go back to our chapter on the sun and its planetary system. Remember, if the sun were more massive, the Earth would revolve around it more quickly. Let's put it in a different way: the Earth has just the right speed to avoid its falling into the sun or flying off into space. Let's put it another way still: we could measure the sun's mass simply by measuring the Earth's speed. I want to emphasize that point. It's going to be important.

Just as the Earth revolves around the sun in one year, and the moon around the earth in one month, the sun itself, along with all the other stars, revolves around the core of our galaxy in about two hundred million years. Knowing the speed of a given star enables us to measure the mass of all the stars and nebulae between it and the centre of the Milky Way. And there we have a problem: all this mass is not enough to keep the stars anchored to the galaxy. The deficit is particularly significant for the stars most distant from the centre. Much more mass would be needed to hold them within the galaxy."

"What does that mean?"

"It means that there is more matter in our galaxy than is visible in the form of stars and nebulae, about six times more! We call that additional invisible matter 'dark matter.'"

"Do we know what it's made of?"

"No. We do know, on the other hand, what it's not made of. It is very different from the matter from which we have been formed. It is not made up of protons, neutrons, electrons, photons, those components of matter that we call 'ordinary.' All the attempts to learn more about it have so far failed." ·

"So what do we know about it?"

"First, that it exists! We now have other proof to confirm it. Its existence was already deduced, around 1935, from the astronomer Fritz Zwicky's observations of galaxy clusters. We also know that, like ordinary matter, it can attract other bodies. In fact, that is what revealed to us that it was there. Another important fact: dark matter represents 24% of all the matter in the cosmos, and ordinary matter, only 4%."

"And the rest? 4% plus 24% makes 28%. That still leaves 72%."

"We'll get back to that soon."

"Couldn't the black holes in our galaxy, that we talked about, make up the dark matter?"

"No, there are not enough of them. All the black holes put together, including that at the centre of the galaxy, would not even make up 1% of the mass needed to keep the stars in place. We can't count on them."

"So what can we count on?"
"On another interesting astronomical discovery of recent years: that of dark energy. We're coming to that now."

Dark energy and the future
of the universe

"What do you mean by dark energy?

"Remember that astronomy is above all an observational field. To understand a term, we must first go back to the observation that gave rise to it.

At the end of the twentieth century, we reasoned in the following way: because of their respective masses, galaxies attract each other mutually. Their movement away from each other must then be slowing down, and they must today be less distant from one another than if no attraction were being exerted on them since the Big Bang. But around 1990, we were astonished to learn that the exact opposite is true. The galaxies are not closer to each other, but farther away from each other than we expected. The measurements are difficult, of course. But they were carefully verified, and now seem very credible.

What is going on? It was concluded that there is another force being exerted on the galaxies. We attribute it to an invisible substance called 'dark energy,' which is present all through the universe. Unlike dark matter and ordinary matter, it does not exert an attraction, but rather a repulsion on everything around it. It accelerates the movement of galaxies, rather than slowing it. As with dark matter, we do not know the nature of this substance. We do know, however, that it represents 72% of cosmic density."

"So let me add it all up: 24% dark matter plus 72% dark energy; that means 96% of the universe is unknown to us. Are we sure that 96% of it is really invisible?"

"In science, absolute certainty does not exist. Let's just say that the arguments in favor of its existence are very convincing. And that makes for a marvelous field of research. Computers do not give us the answer to this question. We must continue to observe and to scratch our heads."

"Grandpa, what can we say about the future of the universe?"

"Now you're telling me to play the prophet. That's a risky game. Most often, prophesies of all kinds have turned out to be false. But it is tempting to try to predict the future on the basis of our current knowledge, the past, and the laws of physics. The first problem comes from the fact that our knowledge of physics is constantly changing. We try to predict the future on the basis of current theories, knowing that new developments could make our claims null and void. But, given these precautions, let us try all the same. Let's look at the Big Bang theory, which is the best description we have for the evolution of the cosmos. Starting with the fact that today galaxies are moving away from each other, what does this theory have to say about the distant future?

Before explaining, I'm going to propose a little experiment. Take a small stone and throw it in the air. We note that it rises up, while progressively slowing down. At a certain height determined by the force with which it was thrown, it stops, changes direction, and comes back to earth, this time accelerating. Do you know what makes the stone slow down as it climbs and accelerate when it falls? It's the force of gravity exerted between it and our planet. We might say that it is trying to escape this force, and to fly off into space. If the initial impetus were adequate (but your arm is not strong enough for that), it would succeed, and would be gone forever. And so there are two possible scenarios for the stone's future: it can fall back to earth or escape into space. And the choice depends on the initial momentum provided by the thrower.

Now let's go back to the expanding universe. The galaxies are moving away from each other, while being mutually attracted at the same time. And there are again two possible scenarios. If the initial impulse at the time of the Big Bang was sufficient (the first scenario), the galaxies would move away from each other forever. As a result, the universe would gradually thin out and cool, until it reached temperatures close to absolute zero. This scenario is called the 'Big Chill.' In the second case, the initial impulse was not strong enough to 'free' the galaxies from their mutual attraction. The temperature of the universe would return to the high values it knew in the past, in a phase called the 'Big Crunch.' Einstein's gravitational theory tells us that both scenarios are, a priori, possible, but it does not tell us which is right. The answer can only come from observing the movement of the galaxies."

'So where does that leave us for predicting the future of the universe?"

"It's the dark energy that seems to hold the key. The problem is that we do not know if it will change over time. If it does not change in the billions of years to come, the expansion will continue to accelerate indefinitely, and we are headed for the Big Chill. On the other hand, if it diminishes, we could heat up on the way to a future Big Crunch. But there is no reason to panic! This won't happen for several dozen billions of years. Where the near future is concerned, the planetary warming we are now experiencing, linked in large part to human activity, is much more unsettling."

"Your answer, in the end, is that we just don't know! The Big Crunch is not out of the question. But if that is so, could it all start over again? A new Big Bang?"

"That's not impossible! Indian myths talk about a universe eternally rising from its ashes, like the phoenix."

"Is that a reason to argue for the Big Crunch scenario?"

"No, it's just an interesting analogy."

Reflections

"I'm looking at the stars, I know how to recognize Arcturus, Altair, Vega, and Deneb, and I always enjoy finding them again. What you've told me makes them all the more fascinating. I would have never guessed that with telescopes we can observe so many galaxies, so many stars, one more extraordinary than the next, and that we can learn so many interesting things about our universe, its past and its history..."

"Our great good fortune is that scientists have spent so many hours creating these instruments, so we can explore the night sky. They have used their observations to confirm or refute the theories that help us to understand what is happening around us. We benefit from the fruits of their labours. Today, still, everywhere in the world, researchers are hard at work trying to solve other mysteries.

There is a very important fact I want to emphasize: it's that we live in a universe that has a history, a universe where new things are always happening that influence what is to come. Here's an example: On February 24, 1987, in the Australian sky, the explosion of a star in the Large Magellanic Cloud was observed with the naked eye. Since that moment, it has propelled into space new atoms that it had been manufacturing all during its lifetime. Another example: a few years ago you were conceived in your mother's belly. And here you are with me looking at the sky and asking me questions...

And so countless events in the heavens and on earth are taking place at every moment of this great saga that I like to call the 'universe adventure.'"

"You mean the adventure of the universe?"

"No, I mean that the universe *is* an adventure! It has been unfolding for nearly fourteen billion years, over a gigantic, perhaps infinite expanse. The sun, our lives, your cat's life, are all short episodes in this epic. It's a succession of linked or juxtaposed events that, in interacting, determine our future direction.

Thanks to astronomy, we have learned that we are not the centre of the world, as had been believed for a long time. The deck chairs from which we look at the sky are set down on a small planet that moves around a yellow star, situated on the edge of a galaxy, one of billions just like it.

Perhaps more astonishing still, our view of time has expanded. For a long time we thought that the world had been created very recently, at most a few thousand years ago. Today we see opening out before us an enormous prospect of billions of years. Our life span which sometimes seems so long is infinitesimal compared to the age of the universe or of the sun. It's like the blink of an eye compared to a year. Mark Twain, a nineteenth century American writer, used another comparison to denounce the vanity of human beings who exaggerated their own importance. The length of our lives is like the thickness of that coat of paint on the top of the Eiffel Tower, compared to the tower's height. At that time, people were unaware of the enormous time span required for life to appear. For the foundations of intelligence to be laid took billions of years, over a physical expanse of tens of billions of light-years. That is another important discovery we owe to scientific research."

"That's marvelous! And yet I know that you're very concerned about what's going to happen next in this beautiful story. What has gone wrong?"

"Yes, we're going to talk about the ecological crisis that now faces us. We can relate it directly to the emergence of intelligence in our human species, to the size of our brains, to its exploits. In this connection, Plato, the Greek philosopher, recounted the following legend: at the birth of the first living people, two brothers, Epimetheus and Prometheus, were given the responsibility for distributing to each species its specific gifts for coping with the dangers in nature. Epimetheus began. He gave memory to the elephants, speed to the felines, flight to the birds. Prometheus saw that his brother had forgotten about humans, and to

117

compensate he endowed them with intelligence. They could then make their own tools or adapt the fires of heaven to their own purposes."

"That's a pretty legend. But in reality, how did it all come about?"

"The appearance of the first humans goes back about two hundred thousand years. At the time, life must not have been easy. The Earth was populated by fearsome predators against which one had to protect oneself and one's children. Human beings were extremely vulnerable in the face of these dangers. They had to eat and try not to be eaten. Intelligence developed as a faculty for competing with other living things. In this sense, we may regard it as a new emergent property that appeared once complexity reached the stage when certain animals came on the scene, and among them, the human species. That is how we made our entry into the ongoing universe adventure. As the ages passed, this faculty, so beneficial at the beginning, began to cause problems. Thanks to it, we human beings developed technologies of tremendous effectiveness. But if on the one hand we invented, for example, excellent medicines, on the other we emptied the oceans, destroyed the forests, and made agricultural lands barren. We eliminated many species of animals and plants that had been on Earth for hundreds of millions of years. We discovered that our planet is not infinite and that we are now face to face with its limits. That is what we now call the ecological

118

crisis. The word ecological means 'that which relates to the house.' We are mistreating our house, the biosphere, and all its inhabitants."

"Might we consider going off to colonize other planets?"

"I don't think that's a good solution. In a short time we would come up against the same limits. It would just be a repetition of what we are doing now, and would only delay the issue."

"Suppose that intelligence were to appear on another planet. Would those who possess it find themselves with the same problems?"

"That is a question we're going to deal with now. To put it in context, we're going to invent a scenario with a lot of 'suppose thats.'

Suppose that life forms, more or less equivalent to those we observe on Earth, had developed on many planets in the universe. Suppose also that their developmental stages were similar to those that occurred here. It's just a hypothesis, but it's instructive. For many billions of years, new stars have continually been formed in galaxies. Some were born long before the sun, which is only 4.5 billion years old; others are much more recent. And so their planetary systems would have different ages. Imagine setting off on a journey to visit different planets. On some, we would find the most primitive of life forms; cells teeming in pools of lukewarm water. Elsewhere, we would see reptiles crossing the savannahs, or the ancestors of birds pollinating the first flowers, or beings gifted with intelligence painting the walls of the grottos where they have taken shelter."

"Could we find biospheres like ours might be in a hundred years, two thousand years, a million years? What state would they be in? That would give us a picture of our own future. Like in a crystal ball!"

"Your question brings us back to our current situation. We can imagine that, just like us, other civilizations faced the same challenges as our own today: coexisting with their own technologies, slowing the deterioration of a biosphere adversely affected by the impact of industry. The ecological crisis we are undergoing could be a universal phenomenon, a predestined phase in growth and complexity everywhere that it reaches such high levels of intelligence and consciousness. A kind of ultimate test to which all the intelligent inhabitants must submit, on planets where life has appeared (or might appear). At stake is the fate of intelligence, its capacity not to disappear along with the species that, having inherited it, did great harm to the biosphere from which it sprang. Our interstellar exploration might reveal a variety of case histories. There where an intelligent species passed its test, the universe adventure continued its evolution towards new heights that we cannot possibly imagine. On the other hand, there where the species failed in its task, we would find the ruins and debris its actions caused. Among these remnants, living things that survived the disaster would be starting over. And if, on Earth, our intelligence were to lead us to a similar situation, the fruits of all our creative facultie the art, the science would be destroyed and soon forgotten. The names of Mozart and Van Gogh would no longer mean a thing. And the admirable fellowship among human beings, their compassion for suffering individuals, would be lost."

"But maybe, after a certain time, a new chapter in evolution would emerge from the ashes of the one before?"

"Yes, you're right. Like many other stars, the sun still has several billion years of life before it. Intelligence would perhaps have another opportunity to flourish, and who knows, to prevail."

"Why not succeed in taking up the challenge right now?"

"That answer is in the hands of our present earthlings."

CONTENTS

Lacrimosa
Régis Jauffret

SALAMMBO
PRESS LIMITED

Lacrimosa unfolds through a moving exchange of letters between the narrator and his young lover, Charlotte, who has just committed suicide. Their poignant dialogue makes this epistolary novel a truly cathartic experience.

'Régis Jauffret has perhaps written his most accomplished novel, a work of devastating and devastated beauty: an ode to a dead lover.'

'*Lacrimosa* works like a literary boxing match, a heartbreaking masterpiece where emotion is never far from the absurd.' **Express**

'Tragic and caustic.'

'A merciless tale in which honesty explodes from every page, sometimes to the point of provocation.' **Figaro**

salammbopress.com

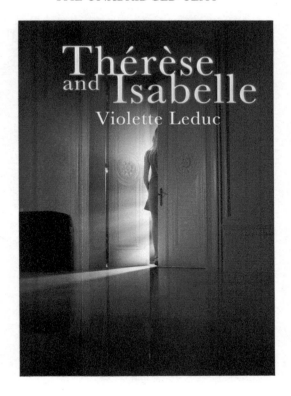

'So here we have extraordinary writing about sex; and, more importantly, about love, and the way it makes us feel.' Nicholas Lezard, *Guardian*

'A classy new translation of Leduc's masterpiece on the tyranny of love.' *Independent*

'Reading Leduc is like discovering a whole new nervous system.' Deborah Levy

'Lesbian story ban is lifted.' *Observer*

salammbopress.com